Contents

1 Introduction to Terminology

Understanding terminology is crucial for clear communication. Precise language prevents ambiguity, ensuring everyone involved shares the same understanding. Misunderstandings, stemming from imprecise wording, can have significant consequences in any field, from a simple miscommunication in a casual conversation to major errors in complex scientific research or legal rulings. Therefore, mastering terminology is not merely about enriching your vocabulary; it's a fundamental skill for effective interaction and problem-solving.

The evolution of terminology reflects the ongoing progress within each field. New discoveries, technological advancements, and evolving social norms constantly introduce new terms or redefine existing ones. Consider the rapid changes in technology; "internet," "website," and "social media" are relatively recent additions to our lexicon, yet they now hold crucial positions in our daily lives. This dynamic nature necessitates continuous learning and adaptation to remain current and effective in your communication. Staying updated involves actively engaging with professional journals, attending conferences, and participating in online communities. This proactive approach ensures that your understanding of terminology remains precise and pertinent.

Effective terminology use contributes to professional credibility. Demonstrating a solid grasp of the specific language of your field shows you understand the intricacies and nuances of your work. This level of expertise builds confidence and trust with colleagues, clients, and other stakeholders. Within a professional setting, the use of appropriate terminology not only elevates your communication but also positions you as a knowledgeable and reliable expert. This is particularly true in specialist fields, where the correct terminology is essential to accuracy and understanding.

Choosing the right word can significantly impact the effectiveness of your communication. Synonyms, while appearing similar, often carry subtle differences in meaning or connotation that can alter the entire message's interpretation. For instance, "cheap" and "inexpensive" may seem interchangeable, but "cheap" often implies low quality, whereas "inexpensive" suggests affordability without compromising quality. This precision in word choice is vital for conveying information accurately and avoiding potentially negative interpretations. Developing a keen awareness of such subtle distinctions allows for nuanced communication, strengthening your ability to express yourself clearly and persuasively.

Precision in terminology is particularly crucial in fields that demand accuracy, such as law, medicine, and science. In law, imprecise language can lead to misinterpretations of contracts or legal documents, resulting in costly mistakes and disputes. Similarly, in medicine, the wrong term can have devastating consequences for patient care. And in science, inaccurate terminology can lead to flawed research and misrepresented findings. In all of these fields, the stakes are high, so the use of precise language isn't just a matter of style; it's a critical factor in ensuring accurate and effective communication, potentially having life-altering implications.

Developing a strong grasp of terminology involves more than just memorizing definitions. It requires a deep understanding of the concepts behind the words. Truly understanding a term means knowing its origins, its nuances, its relationship to other terms in the field, and the contexts in which it is appropriately used. This comprehensive approach facilitates not just the correct usage of words but also a deeper comprehension of the subject matter itself, which subsequently informs better communication and clearer expression of complex ideas. This goes beyond simply knowing the definition; it's about internalizing the underlying meaning and application of the term.

The effective learning of specialized terminology requires consistent effort and strategic engagement with relevant materials. Passive reading alone is rarely sufficient. Active learning strategies, such as creating flashcards, participating in discussions, and applying the terminology in written and spoken exercises, significantly improve retention and understanding. Making the learning process active and interactive ensures that you are not just passively absorbing information but actively engaging with the material, building a stronger, more reliable foundation of knowledge. This includes using the terminology in practice, either through writing or speaking exercises.

Access to reliable resources is crucial for building a comprehensive understanding of terminology. High-quality dictionaries, encyclopedias, and specialized textbooks are invaluable tools for learning definitions and exploring the nuances of language within a specific field. However, the reliability of the source must be carefully considered, as inaccuracies can lead to incorrect understanding and usage of terminology. It is always best to consult multiple reliable sources to confirm the accuracy of the information obtained. Such a multi-faceted approach contributes to acquiring well-grounded knowledge, minimizing the risk of misconceptions.

1.1 Defining Terminology

Precise language is the cornerstone of clear communication, particularly in specialized fields. Ambiguity is the enemy of accuracy, leading to misunderstandings, misinterpretations, and, in some cases, serious consequences. Consider the legal profession, where the wrong word can alter the meaning of a contract or impact the outcome of a trial. The same holds true for medical diagnoses, scientific research, and technical specifications. A single poorly chosen term can derail an entire project, leading to wasted time, resources, and potentially, significant financial loss. Mastering precise terminology isn't just about memorizing definitions; it's about understanding the nuances of language and its impact within a specific context.

This understanding extends beyond simply knowing the definition of a word. It involves grasping the word's historical context, its etymology, and its various connotations. For example, the word "algorithm" in computer science has a precise meaning, referring to a step-by-step procedure for solving a problem. However, the word's use in everyday conversation might be more casual, often implying a general method or procedure. This subtle difference highlights the importance of understanding the specific meaning within the field of study. The successful professional appreciates these subtleties, using language as a precise tool for effective communication. Their expertise isn't just limited to the subject matter; it also includes the mastery of its corresponding vocabulary.

Let's consider the legal field as a prime example. In contract law, the terms "offer," "acceptance," and "consideration" have very specific legal meanings, forming the basis of a legally binding agreement. Misunderstanding these terms can lead to disputes, protracted negotiations, and costly litigation. Similarly, in criminal law, the distinction between "mens rea" (guilty mind) and "actus reus" (guilty act) is crucial in determining criminal liability. A legal professional's fluency in these precise terms is essential, ensuring accurate interpretation of laws and effective advocacy for their clients. In essence, precision in legal terminology protects rights and ensures justice.

The scientific realm, particularly physics, demands extreme precision. Consider the difference between "mass" and "weight." While often used interchangeably in casual conversation, in physics, they represent distinct concepts: mass refers to the amount of matter in an object, while weight is the force exerted on that object by gravity. A physicist must understand this distinction to correctly apply the laws of physics. Using them imprecisely can lead to inaccurate calculations, flawed theories, and ultimately, a lack of progress in the field. Understanding terminology therefore facilitates progress and advances knowledge, both for the individual and for the collective scientific community.

Moving into the technological sphere, imagine the challenges of building complex software systems without a precise understanding of terms like "algorithm," "data structure," or "object-oriented programming." These terms are not mere jargon; they represent fundamental concepts essential to the design, development, and maintenance of software. In the absence of precision, software engineers risk creating systems that are inefficient, unreliable, and prone to errors. The consequences can range from minor inconveniences to major security breaches, with significant financial and reputational risks. Clear and precise communication through accurate terminology is paramount for smooth and successful collaboration, across different teams and companies.

The medical profession offers perhaps the clearest example of the life-or-death importance of accurate terminology. A misdiagnosis, stemming from imprecise communication, can have devastating consequences. Doctors, nurses, and other healthcare professionals must use medical terminology with absolute precision to ensure patient safety and optimal treatment. The precise naming of diseases, the detailed description of symptoms, and the clear explanation of treatment plans are all vital components of effective medical practice. Consider the potential risks in misinterpreting a patient's chart due to ambiguous medical terminology, impacting the entire healthcare treatment plan.

This importance of precise terminology extends across all disciplines. From the intricacies of legal language to the exacting demands of scientific research, and the critical precision demanded by medical practice, mastering the terminology of a given field is indispensable to success. This chapter has only scratched the surface, but the underlying principle remains: mastering terminology is not merely an academic exercise but a crucial skill for professional competence and effective communication. The use of precise language ensures clarity, reduces ambiguity, and ultimately leads to better outcomes in any given field. Therefore, actively developing your vocabulary, within your chosen subject area, is an ongoing and crucial aspect of professional development. It is an investment that will continually pay dividends.

1.2 The Importance of Precision

Precision isn't merely about choosing the right word; it's about understanding the nuanced implications of each term. A single word, subtly misplaced, can alter the entire meaning of a sentence, potentially leading to misunderstandings with significant consequences, especially within specialized fields. This is particularly true in legal contexts, where a misplaced adjective or a poorly defined clause can invalidate a contract or alter a legal judgment. Consider, for example, the difference between "negligence" and "gross negligence" in a civil lawsuit. The former implies a careless mistake, while the latter suggests a reckless disregard for consequences, leading to drastically different levels of liability. The precision demanded by the law underscores the critical need for accurate terminology in all professional fields. This is not just about legal nuances; the requirement for specificity extends to every area where accurate communication is vital.

The importance of precision extends beyond legal ramifications, impacting technical fields significantly. Think of the healthcare sector; miscommunication in a surgical procedure or a medication prescription can have devastating effects. Precise terminology is essential not only for the communication between healthcare professionals, but also for ensuring patient understanding and compliance. Imagine the repercussions of an inaccurate diagnosis or a misunderstood treatment plan. The consequences are severe and far-reaching, demonstrating the vital necessity of unambiguous language. The accuracy needed in healthcare necessitates a deep understanding of medical terminology, highlighting the significance of this book's focus on precision in language.

Even in seemingly less critical contexts, imprecision can lead to costly errors. In business, for instance, inaccurate financial reporting or a flawed marketing strategy, stemming from vague terminology, can severely harm a company's performance. A misinterpretation of a contract clause can result in financial losses, while a poorly defined project scope can lead to project delays and budget overruns. Thus, even seemingly minor imprecisions in business terminology can have significant consequences. The use of precise language in all business communications is not just a matter of style, it's an essential element of effective management and sustainable growth.

The pursuit of precision requires a dedicated effort. It means going beyond surface-level understanding and delving into the etymology and context of each term. We must strive to understand not only the denotation (the literal meaning) but also the connotation (the implied meaning) of the words we use. For example, "innovation" might conjure up images of groundbreaking technology in one context and incremental improvement in another. These subtle differences can profoundly impact the interpretation of any given statement, underscoring the importance of precise word choice in all fields. This means consistently questioning our own assumptions and seeking clarification when necessary.

Furthermore, the development of specialized terminology often reflects the evolution of a field itself. As scientific understanding advances, for example, new terms emerge to describe previously unknown phenomena, or old terms gain new and more precise definitions. For instance, consider how the understanding of "gravity" has developed from a simple descriptive concept to a highly sophisticated theory of general relativity. This continuous refinement of terminology mirrors the ongoing evolution of knowledge, adding another layer to the importance of continuous learning and precise language use. Such precision allows for clearer, more efficient communication within a field, leading to better understanding and accelerated progress.

This book aims to equip readers with the tools they need to navigate the complexities of different terminologies with confidence and accuracy. By providing comprehensive definitions, clarifying nuances, and highlighting potential pitfalls, this compendium hopes to foster a deeper appreciation for the critical role precision plays in all forms of professional communication. Remember, this is not just about memorizing definitions; it's about cultivating a mindset of careful consideration, a dedication to clarity, and an understanding of the profound impact that precise language has in shaping our world. Achieving true mastery requires continuous effort, and this book serves as a valuable resource on that journey. Mastering precise communication is a journey, not a destination, and consistent practice, combined with this book's resources, will empower the readers to speak and write with greater confidence and understanding.

2 Science Terminology

Understanding scientific terminology is crucial for anyone seeking to engage with scientific literature or discussions. It's a language of precision, built on rigorously defined terms, many with Greek or Latin roots revealing their meaning. A grasp of these terms unlocks a deeper comprehension of complex concepts, moving beyond vague understanding to a solid, nuanced grasp of the subject matter. Without this foundation, even the most lucid explanations can remain opaque.

Physics, for example, relies heavily on specific terminology to describe fundamental forces and interactions. Concepts like "Newton's Laws of Motion," encompassing inertia, acceleration, and the action-reaction principle, are cornerstones of classical mechanics, but understanding their precise implications requires familiarity with related terminology such as "force," "mass," "velocity," and "momentum." Delving into the realm of electromagnetism involves grasping terms like "electric field," "magnetic field," "voltage," and "current." Each term represents a specific phenomenon, and understanding their interrelationships is key to understanding the behavior of electricity and magnetism. Furthermore, thermodynamics introduces concepts like "entropy," "enthalpy," and "Gibbs free energy," demanding careful definition and application.

Chemistry utilizes a distinct vocabulary to describe the composition and behavior of matter. Understanding atomic structure requires a firm grasp of terms like "atom," "electron," "proton," and "neutron." Similarly, comprehending chemical reactions necessitates familiarity with "reactants," "products," "stoichiometry," and "equilibrium." Organic chemistry, a vast field in itself, employs terms like "isomerism," "functional groups," and "stereochemistry" to classify and differentiate organic compounds, each term carrying its own weight and significance. The vocabulary is extensive, demanding both rote memorization and an understanding of underlying concepts.

Biology's terminology is equally rich and varied, spanning diverse areas like genetics, cellular biology, and ecology. Genetics hinges on understanding terms like "gene," "chromosome," "DNA," and "RNA," each term defining a crucial element in the process of heredity and genetic expression. Cellular biology is another complex field, necessitating knowledge of terms such as "mitosis," "meiosis," "endoplasmic reticulum," and "Golgi apparatus," each referencing specific cellular structures and functions. Ecology, the study of interactions between organisms and their environment, introduces terms such as "biodiversity," "ecosystem," "trophic levels," and "niche," revealing the complexity of the interconnected web of life.

It's not simply about memorizing isolated definitions. The real challenge lies in understanding the relationships between terms, how they function within larger conceptual frameworks. For instance, in understanding concepts of energy transfer in biological systems, you must link terms such as "ATP," "cellular respiration," and "photosynthesis," recognizing how these processes interrelate and depend on one another. To properly understand evolution, the concepts of "natural selection," "adaptation," "speciation," and "genetic drift" must be understood in relation to one another. It is this intricate web of interconnectivity that forms the complex tapestry of scientific understanding.

Therefore, the effective study of scientific terminology involves a multi-faceted approach. Start with precise definitions, using reputable scientific dictionaries and textbooks. Then, actively seek to connect these definitions to real-world examples and illustrations, ensuring your understanding is robust and adaptable. The aim is not simply rote learning, but a genuine comprehension of the underlying scientific principles. Only then does the intricate language of science begin to truly reveal its secrets. Active engagement, involving problem-solving and application of concepts, is vital for deeper understanding, ensuring that this specialized vocabulary is not merely memorized but truly integrated into a meaningful scientific framework. This, ultimately, is the path to fluency in the language of science.

2.1 Physics Terms

Classical mechanics introduces concepts like velocity, acceleration, and force, describing the motion of objects. Newton's laws of motion, fundamental to this branch, dictate the relationship between a body's mass, its acceleration, and the net force acting upon it. Understanding these laws requires grasping the meaning of terms such as inertia, momentum, and work. The application of these principles extends to various areas, including projectile motion, rotational motion, and the study of simple harmonic oscillators. Grasping the nuances of these terms is crucial for comprehending the behavior of objects in everyday life as well as in more complex physical systems.

Thermodynamics delves into the relationship between heat, work, and energy. Key terms in this field include temperature, heat capacity, and entropy. The laws of thermodynamics—the conservation of energy, the increase of entropy, and the unattainability of absolute zero—are fundamental principles governing energy transfer and transformations. Understanding these laws requires familiarity with concepts like internal energy, enthalpy, and Gibbs free energy. Applications of thermodynamics span from the efficiency of engines to the spontaneity of chemical reactions. Mastering these concepts is fundamental to various engineering disciplines, and it helps in comprehending processes such as energy production and heat transfer.

Electromagnetism explores the interaction between electric and magnetic fields. Fundamental concepts include electric charge, electric field, magnetic field, and electromagnetic radiation. Coulomb's law describes the force between charges, while Faraday's law and Ampere's law govern the relationships between changing electric and magnetic fields. The understanding of these laws is vital to interpreting phenomena like electromagnetic induction and the propagation of light. The concepts of electric potential, capacitance, and inductance are also essential, forming the basis of numerous technological applications, from electric circuits to power generation. These terms are also fundamental to modern technologies and the understanding of fundamental physics.

Modern physics introduces concepts that extend beyond classical mechanics and electromagnetism. Relativity, pioneered by Einstein, revolutionized our understanding of space, time, and gravity. Key terms include spacetime, time dilation, and length contraction, and understanding them requires grasping the principles of special and general relativity. Quantum mechanics, another cornerstone of modern physics, describes the behavior of matter at the atomic and subatomic levels. Crucial terms here include wave-particle duality, quantization, and the uncertainty principle. These concepts are challenging but essential for comprehending the nature of matter and energy at the fundamental level, shaping our technological progress in fields like electronics and materials science. Mastering these terms is vital to understanding modern physics.

Optics, focusing on the behavior and properties of light, involves terms like reflection, refraction, diffraction, and interference. Snell's law describes the refraction of light at interfaces between different media. The wave nature of light is central to understanding diffraction and interference phenomena. Concepts such as polarization, lenses, and optical instruments are also critical in optics and applied optics. These terms are applied in many areas, from designing optical systems to understanding the nature of light itself. Understanding the optics terms is vital in many applications.

Nuclear physics explores the properties and behavior of atomic nuclei. Key terms include radioactivity, nuclear fission, and nuclear fusion. Understanding isotopes and nuclear reactions requires familiarity with concepts like nuclear binding energy and half-life. These processes are fundamental to nuclear power generation and medical imaging techniques. Mastering these terms can significantly assist in understanding some technological advancement and applications.

Finally, particle physics investigates the fundamental constituents of matter and their interactions. Terms like quarks, leptons, and bosons are crucial for understanding the standard model of particle physics. Concepts such as the strong force, the weak force, and the electromagnetic force govern the interactions between these particles. The discovery and characterization of new particles continually refine our understanding of the universe's fundamental building blocks. Understanding the terminology of particle physics is fundamental to interpreting findings in high-energy experiments and expanding the model that describes the universe.

2.2 Chemistry Terms

Atoms, the fundamental building blocks of matter, are composed of a nucleus containing protons and neutrons, orbited by electrons. The number of protons defines the element's atomic number, determining its position on the periodic table and its chemical properties. Isotopes are atoms of the same element with varying numbers of neutrons, affecting their mass but not their chemical behavior. Understanding these basic concepts is crucial for comprehending more complex chemical phenomena. Learning this is the foundation of all advanced studies in chemistry. Ions are atoms that have gained or lost electrons, resulting in a net electric charge. This charge influences their interactions with other atoms and molecules. The concept of electronegativity, a measure of an atom's tendency to attract electrons in a chemical bond, dictates the type of bond formed – covalent (shared electrons) or ionic (electron transfer). Mastering the intricacies of these foundational ideas paves the way for deeper exploration of chemistry.

Chemical reactions involve the rearrangement of atoms and molecules to form new substances. They are represented by balanced chemical equations, where reactants are transformed into products, conserving mass and charge. Stoichiometry deals with the quantitative relationships between reactants and products, allowing chemists to predict the amount of product formed from a given amount of reactants. Reaction rates, influenced by factors like temperature and concentration, determine the speed at which these transformations occur, and are critical to understanding dynamic equilibrium and reaction mechanisms. Catalysis, which accelerates reactions without being consumed, exemplifies how seemingly small changes can have large impacts. The understanding of these factors is critical to industrial chemical processes.

Acids and bases are substances that donate or accept protons, respectively, according to the Brønsted-Lowry definition. The pH scale measures the concentration of hydrogen ions, reflecting the acidity or alkalinity of a solution. Neutralization reactions involve the combination of an acid and a base to produce water and a salt. Buffers are solutions that resist changes in pH upon the addition of small amounts of acid or base, maintaining a stable environment crucial for many biological and chemical processes. Acidity and basicity impact a vast range of chemical reactions, from industrial processes to biological functions. This seemingly simple concept has wide reaching consequences, including its effect on the environment and other areas of applied chemistry.

Organic chemistry focuses on carbon-containing compounds, the basis of life itself. Hydrocarbons, composed of only carbon and hydrogen, are the simplest organic molecules, forming the foundation for more complex structures. Functional groups, specific atoms or groups of atoms, impart characteristic chemical properties to organic molecules. Isomerism, where molecules have the same molecular formula but different structural arrangements, results in different properties and functionalities. Stereochemistry, the study of the three-dimensional arrangement of atoms in molecules, is particularly crucial in understanding biological molecules and their interactions. The complexity of organic chemistry's terminology can be daunting, but it is worth mastering it.

Nomenclature in chemistry follows systematic rules to name compounds, avoiding ambiguity. For example, the IUPAC (International Union of Pure and Applied Chemistry) system provides standardized naming conventions for organic and inorganic compounds. Knowing these naming conventions helps avoid confusion when referencing compounds. Familiarity with these rules is crucial for effective communication between chemists and in the interpretation of chemical literature. Therefore, understanding this is paramount to becoming a knowledgeable chemist. It's one of the cornerstones for any aspiring chemist. The systematic names and structural formulas communicate information efficiently and accurately. This ability to accurately communicate and understand chemical data is paramount to efficient collaboration.

Spectroscopy, a powerful analytical technique, involves studying the interaction of matter with electromagnetic radiation. Different spectroscopic methods, such as infrared (IR), nuclear magnetic resonance (NMR), and mass spectrometry (MS), provide valuable information about the structure and composition of molecules. The interpretation of the resulting spectra requires specific expertise and is a critical tool in chemical analysis. These analytical techniques are crucial in many scientific fields, not just chemistry. This is often a pivotal point of study and research in many scientific endeavors. The combined use of several techniques allows for accurate structural elucidation. The information derived aids in the identification of unknown compounds and for quality control of chemical products.

Thermochemistry examines the heat changes accompanying chemical reactions. Enthalpy, the heat content of a system at constant pressure, indicates whether a reaction releases (exothermic) or absorbs (endothermic) heat. Entropy, a measure of disorder in a system, influences the spontaneity of reactions. Gibbs free energy combines enthalpy and entropy to predict the spontaneity and equilibrium constant of a reaction. These thermodynamic principles underpin many chemical processes and are integral to understanding reaction feasibility. The ability to predict the energy changes helps in designing chemical processes more efficiently, reducing energy consumption. Understanding these fundamentals and using these principles ensures sustainability and efficiency. It's a key component in green chemistry and in reducing the carbon footprint of chemical industries.

Finally, remember to consult reliable chemical dictionaries and handbooks for additional terminology. The field of chemistry is vast and constantly evolving, with new terms and concepts continually emerging. Staying updated with the latest terminology is essential for any serious chemistry student or professional. It is critical to use the right terminology and stay up-to-date. The use of precise, accurate terminology is essential for clear communication and understanding, and therefore crucial for progress in scientific endeavors. Constant learning and refinement of knowledge are essential for staying current in a field as dynamic and ever-evolving as chemistry. The value of updated information cannot be overstated. Consistent review and updating of your knowledge base is critical for success.

2.3 Biology Terms

Cellular processes are fundamental to understanding life. Metabolism, the sum of all chemical reactions within an organism, involves catabolism (breaking down complex molecules) and anabolism (building up complex molecules). These intricate pathways are regulated by enzymes, biological catalysts that speed up reactions. Understanding metabolic pathways is crucial for comprehending how organisms obtain and utilize energy, a key concept in biology. The processes of photosynthesis, where plants convert light energy into chemical energy, and cellular respiration, where organisms release energy from glucose, are prime examples of metabolic processes with wide-reaching consequences for ecosystems. Furthermore, the detailed study of individual metabolic pathways, their intricate regulation mechanisms, and their connections with other cellular processes provides vital insights into the workings of life itself. Understanding even a single metabolic pathway requires a deep dive into biochemistry, including enzyme kinetics, reaction mechanisms, and regulatory controls. For instance, the citric acid cycle or Krebs cycle, a central metabolic pathway in aerobic respiration, is a complex network of reactions involving numerous enzymes and intermediates, the detailed understanding of which is essential for grasping cellular energy production.

Genetics, the study of heredity, is built upon the structure and function of DNA (deoxyribonucleic acid), the molecule that carries genetic information. DNA replication, the process by which DNA makes copies of itself, is essential for cell division and inheritance. This process involves unwinding the DNA double helix, synthesizing new strands complementary to the existing ones, and proofreading to minimize errors. Understanding the mechanics of DNA replication requires knowledge of enzymes like DNA polymerase and helicase, and their roles in the process, also providing insights into the fidelity of DNA replication and the mechanisms that maintain the integrity of the genome. Mutations, changes in the DNA sequence, can lead to variations in traits and may drive evolutionary change, or cause genetic diseases. The study of these mutations and their effects lies at the heart of understanding the evolutionary processes and the origins of genetic diversity. The detailed mechanisms of DNA repair, a critical process that counteracts the damaging effects of mutations, are also of paramount importance in comprehending genomic stability and the processes of genetic inheritance and variation.

Evolutionary biology explains the diversity of life on Earth. Natural selection, the process by which organisms with advantageous traits are more likely to survive and reproduce, is a driving force of evolution. Understanding natural selection involves exploring concepts like adaptation, speciation, and phylogenetic trees, which illustrate the evolutionary relationships between different species. The study of fossils provides crucial evidence of past life forms and their evolution, allowing scientists to reconstruct evolutionary history and understand the transition of life forms throughout time. The concept of punctuated equilibrium, explaining how species may show relatively little change for long periods and then undergo rapid evolutionary change, adds another layer of complexity to the overall picture of evolution. Furthermore, understanding the impact of environmental factors, geographical isolation, and genetic drift on evolutionary trajectories is critical in building a comprehensive understanding of evolutionary mechanisms and principles.

Ecology studies the interactions between organisms and their environment. Ecosystems comprise communities of organisms and their physical surroundings, interconnected through complex food webs and nutrient cycles. Biodiversity, the variety of life within an ecosystem, is essential for its stability and resilience. Understanding ecological concepts such as trophic levels, niche partitioning, and keystone species is crucial for comprehending how ecosystems function and how human activities may impact them. Factors like climate change, habitat loss, and pollution pose significant threats to biodiversity and ecosystem health; therefore, the precise terminology related to climate change impacts, habitat fragmentation, and pollution's effects on ecosystems is necessary for effective conservation and management efforts. The application of ecological principles informs environmental policies and conservation strategies, aiming to protect biodiversity and maintain the health of natural systems for future generations.

3 Technology Terminology

Software terms encompass a vast landscape, from the intricacies of programming languages like Python and Java, their syntax, and functionalities to the abstract concepts of algorithms and data structures. Understanding terms like "object-oriented programming," "recursion," and "polymorphism" is crucial for anyone working in software development. Moreover, grasping the nuances of different database management systems, including SQL and NoSQL databases, and their associated terms like "normalization," "transactions," and "indexing" is vital. The ever-evolving world of software demands constant learning, a commitment to staying updated on the latest frameworks and libraries, and the ability to decipher technical documentation. It requires deep understanding of the architecture of applications, the concepts of front-end and back-end development, and the interplay between user interfaces and data processing.
Hardware terms, on the other hand, deal with the physical components of computer systems. This includes understanding the inner workings of processors, their clock speeds, and cache sizes. Terms like "multi-core," "hyperthreading," and "instruction set architecture" are essential for those interested in system performance and optimization. Similarly, knowledge of memory types, such as RAM and ROM, their capacities, and access speeds is necessary. Understanding peripheral devices, including input devices like keyboards and mice, output devices like monitors and printers, and storage devices like hard drives and SSDs,

along with their respective interfaces and functionalities is equally important. Finally, grasping the concepts of system buses, their bandwidth, and the impact on overall system performance is a crucial aspect of hardware understanding.

Networking terminology centers on the communication aspects of computer systems. Fundamental concepts such as IP addresses, subnet masks, and routing protocols—like TCP/IP, BGP, and OSPF—are essential for comprehending network architecture and functionality. Understanding different network topologies, including star, bus, ring, and mesh networks, along with their strengths and weaknesses, is vital. Beyond the basics, concepts like network security, firewalls, virtual private networks (VPNs), and intrusion detection systems play a critical role in securing network infrastructures. Furthermore, the rise of cloud computing introduces terms such as IaaS, PaaS, and SaaS, alongside concepts of virtualization, containers, and serverless architectures. Mastering these concepts unlocks a deeper comprehension of modern network infrastructure and management.

Cybersecurity terms are critical in today's digital world, given the constant threat of cyberattacks. Understanding core concepts such as vulnerabilities, exploits, and malware is crucial for building and maintaining secure systems. Terms like "phishing," "ransomware," "denial-of-service attacks," and "social engineering" represent the diverse range of threats that require careful consideration and proactive mitigation strategies. On the other hand, protective measures like firewalls, intrusion detection systems, antivirus software, and encryption techniques are essential for safeguarding sensitive data and ensuring system integrity. Furthermore, the concepts of risk assessment, incident response, and security auditing are integral parts of establishing and maintaining a robust cybersecurity posture. The ever-evolving nature of cyber threats demands continuous learning and adaptation to the latest techniques and technologies.

The field of AI (artificial intelligence) introduces a unique set of terminology, encompassing the processes and methodologies behind intelligent systems. Fundamental concepts such as machine learning, deep learning, and neural networks provide the foundation for understanding how AI systems learn and make decisions. Terms such as "supervised learning," "unsupervised learning," "reinforcement learning," and "deep reinforcement learning" represent the various learning paradigms used in developing AI systems. Further terminology includes terms like "artificial neural networks," "convolutional neural networks," "recurrent neural networks," and "generative adversarial networks," reflecting the diverse architectures employed in building sophisticated AI models. Finally, understanding the ethical considerations of AI development and deployment, along with the challenges around bias, transparency, and accountability, completes a holistic understanding of this rapidly growing field.

3.1 Software Terms

Software development relies heavily on precise terminology. Understanding these terms is crucial, not just for programmers but for anyone interacting with technology. Misunderstanding a single term can lead to significant project delays, miscommunication, and even software failure. Mastering the language of software development is akin to learning a new dialect, one that blends mathematical precision with creative problem-solving.

Consider, for example, the term "algorithm." It's not simply a set of instructions; it's a precisely defined sequence of steps designed to solve a specific computational problem. The efficiency of an algorithm is paramount, as it directly impacts the speed and resource consumption of the software. We evaluate algorithms based on their time and space complexity, factors that dictate how long they take to execute and how much memory they require. These complexities are expressed using Big O notation, a mathematical shorthand that helps developers compare the efficiency of different algorithms.

Then there are "data structures," which are ways of organizing and storing data in a computer's memory. Choosing the correct data structure is as important as selecting the right algorithm. Arrays, linked lists, trees, and graphs are just a few examples, each optimized for different types of data and operations. A poorly chosen data structure can lead to significant performance bottlenecks. Understanding the strengths and weaknesses of different data structures allows developers to make informed decisions that maximize software efficiency.

Moving beyond the foundational concepts, we encounter terms related to software development methodologies. "Agile" development, for instance, emphasizes iterative development, frequent feedback, and adaptability to changing requirements. In contrast, "waterfall" methodologies follow a more linear, sequential approach. Each has its strengths and weaknesses, and the choice depends on the project's nature and scale. Understanding these contrasting approaches allows for more informed project management decisions.

Software testing is another area rich in specialized terminology. "Unit testing" focuses on individual components of the software, while "integration testing" examines the interaction between different modules. "System testing" verifies the entire system's functionality, and "user acceptance testing" assesses whether the software meets the end-users' needs. Comprehensive testing using a variety of methods is crucial for ensuring software quality and reliability. The thoroughness of testing directly impacts the software's stability and resilience against unexpected issues.

Furthermore, the software development lifecycle (SDLC) involves numerous stages, each with its own vocabulary. From requirements gathering and design to implementation, testing, and deployment, each phase requires a specific set of skills and techniques. Understanding the SDLC helps one to comprehend the software development process as a whole. Consider terms like "requirements elicitation," "design patterns," "code refactoring," and "deployment

pipelines"—each vital for a successful software project.
Finally, let's delve into the specifics of programming languages. Each language has its own syntax, semantics, and paradigms. Object-oriented programming, procedural programming, and functional programming represent distinct approaches to software design. Understanding these paradigms empowers developers to choose the language and approach best suited for a particular project. A grasp of concepts like inheritance, polymorphism, and encapsulation is essential for understanding object-oriented languages, whilst the functional paradigm demands familiarity with concepts like immutability and lambda functions. The selection of an appropriate programming language depends heavily on project requirements, team expertise, and the existing software infrastructure. This underscores the importance of understanding the nuances of different programming paradigms and languages.

3.2 Hardware Terms

Central Processing Unit (CPU), often called the processor or microprocessor, is the brain of your computer. It executes instructions from software, performing calculations and controlling the flow of data. Think of it as the conductor of an orchestra, coordinating all the different parts of the system to work together seamlessly. Understanding its clock speed, measured in gigahertz (GHz), gives you an indication of how many processing cycles it can complete per second – higher GHz generally means faster processing. The number of cores, essentially independent processing units within the CPU, determines its ability to handle multiple tasks simultaneously. More cores usually translate to better multitasking capabilities, crucial for modern applications demanding high performance. However, the architecture of the CPU, encompassing aspects like instruction set and cache size, also significantly impacts performance. Cache, a small, fast memory located within the CPU, stores frequently accessed data for quicker retrieval, substantially improving processing speed. Therefore, focusing solely on GHz or core count provides an incomplete picture; a comprehensive understanding requires delving into the intricacies of CPU architecture. Factors like TDP (Thermal Design Power), indicating the amount of heat the CPU generates, also influence your hardware choices.

Random Access Memory (RAM) is your computer's short-term memory. It's where the operating system, applications, and data currently in use are stored for rapid access. Unlike hard drives or solid-state drives (SSDs), which provide long-term storage, RAM is volatile, meaning data is lost when the power is turned off. The amount of RAM, measured in gigabytes (GB), directly impacts your system's responsiveness. Insufficient RAM can lead to sluggish performance, frequent crashes, and the infamous "out of memory" errors. Choosing the right amount of RAM depends on your usage; intensive tasks like video editing or gaming require significantly more RAM than basic tasks like word processing. The speed of RAM, measured in MHz (megahertz), affects how quickly data can be accessed. Faster RAM generally means quicker application loading times and improved overall system responsiveness, although the gains from increasing RAM speed are often less impactful than increasing RAM capacity. Different types of RAM exist, such as DDR3, DDR4, and DDR5, each having different speeds and power consumption characteristics. Ensure compatibility with your motherboard when selecting RAM.

Storage devices, such as hard disk drives (HDDs) and solid-state drives (SSDs), serve as your computer's long-term memory. HDDs use spinning platters and read/write heads to store data magnetically, while SSDs use flash memory chips, providing significantly faster access times and improved durability. HDDs are generally cheaper per gigabyte, but SSDs offer superior performance. The capacity of storage devices, measured in gigabytes (GB) or terabytes (TB), dictates how much data you can store. Choosing the right storage capacity depends on your needs, considering factors like the size of your operating system, applications, and personal files. Data transfer rates, measured in MB/s (megabytes per second) or GB/s (gigabytes per second), indicate how fast data can be read from and written to the storage device. Faster data transfer rates lead to quicker application loading times and file transfers, improving overall system responsiveness. The interface used (SATA, NVMe, etc.) influences transfer speeds, with NVMe (Non-Volatile Memory Express) generally offering much faster speeds than SATA.

Graphics Processing Unit (GPU), commonly known as a graphics card, is a specialized electronic circuit designed to rapidly manipulate and alter memory to accelerate the creation of images in a frame buffer intended for output to a display device. In simpler terms, it handles the rendering of images and videos on your screen. It's particularly important for tasks like gaming, video editing, and 3D modeling. Key specifications include the GPU model (e.g., NVIDIA GeForce RTX 3060, AMD Radeon RX 6600), memory capacity (VRAM), and clock speed. Higher clock speeds and larger VRAM capacities lead to improved graphics performance and higher frame rates in games. Consider the power consumption of the GPU, as high-end GPUs require more power and potentially stronger power supplies. The type of video output interfaces (e.g., HDMI, DisplayPort) influence connectivity options with your monitor. Choosing a GPU requires carefully considering your needs and budget; high-end GPUs offer significant performance boosts but command higher prices.

Motherboard, sometimes called the mainboard, is the central printed circuit board (PCB) in a computer that connects all the components together. It acts as a communication hub, allowing the CPU, RAM, storage devices, and other components to interact with each other. Choosing a motherboard requires considering the CPU socket type, as it must be compatible with your chosen processor. The chipset determines the capabilities of the motherboard, including the number and type of expansion slots, supporting different technologies like USB, SATA, and PCIe. Memory slots determine how much RAM your motherboard can support. Expansion slots, such as PCIe slots, allow for the addition of graphics cards, sound cards, and other expansion boards. The form factor of the motherboard (e.g., ATX, Micro-ATX, Mini-ITX) determines its size and compatibility with your computer case. Consider features like built-in Wi-Fi, Bluetooth, and audio capabilities when selecting a motherboard. The motherboard's BIOS (Basic Input/Output System) is crucial for booting up your computer. Choosing a motherboard involves navigating a range of specifications crucial for overall system compatibility and performance.

3.3 Networking Terms

Network topologies describe the physical or logical layout of a network. Common topologies include bus, star, ring, mesh, and tree. Understanding these is crucial for designing efficient and reliable networks; each offers different advantages and disadvantages regarding scalability, performance, and fault tolerance. A bus topology, for instance, is simple but suffers from a single point of failure, while a star topology, commonly used in home and small office networks, provides better fault tolerance. More complex topologies like mesh networks are used in larger, more robust systems, offering high redundancy and bandwidth. Choosing the right topology depends heavily on the size, requirements, and budget of the network.

Routers are network devices that forward data packets between networks. They operate at the network layer (Layer 3) of the OSI model, using IP addresses to determine the best path for data to reach its destination. Routers are essential for connecting different networks, such as the internet and a local area network (LAN). They use routing protocols, such as RIP and OSPF, to exchange routing information and build routing tables. Understanding routing protocols is critical for network administrators, enabling them to configure and troubleshoot complex network setups efficiently. Efficient routing is key to optimized network performance.

IP addresses are numerical labels assigned to each device connected to a network that uses the Internet Protocol for communication. They are fundamental for addressing and routing data packets across networks. IPv4 addresses, the older standard, are 32-bit addresses represented in dotted decimal notation (e.g., 192.168.1.1). IPv6 addresses, a newer standard designed to address the depletion of IPv4 addresses, are 128-bit addresses represented in hexadecimal notation. Understanding IP addresses is vital for network configuration, troubleshooting, and security. Incorrect IP addressing can lead to connectivity issues and network failures.

Subnetting divides a larger network into smaller, more manageable subnetworks. This improves network efficiency and security by reducing broadcast domains and facilitating better traffic management. Subnetting involves assigning a subnet mask to each subnetwork, defining which part of the IP address identifies the network and which part identifies the host. Proper subnetting is crucial for optimizing network performance and scalability, particularly in larger networks. Incorrect subnetting can lead to routing problems and connectivity issues.

Network protocols define the rules and standards for communication between network devices. Examples include TCP/IP, HTTP, FTP, and DNS. TCP/IP, the foundation of the internet, provides a reliable, ordered stream of data transmission. HTTP is used for transferring web pages. FTP handles file transfers. DNS translates domain names into IP addresses, making it easier for users to access websites without needing to remember IP addresses. Understanding network protocols is important for network designers, administrators, and developers. A solid grasp of network protocols is crucial for understanding and troubleshooting network-related problems.

MAC addresses, or Media Access Control addresses, are unique physical addresses assigned to network interface cards (NICs). Unlike IP addresses, which can change, MAC addresses are permanently assigned to the hardware. They are used at the data link layer (Layer 2) of the OSI model for local communication on a network segment. MAC addresses are crucial for network security, enabling network devices to identify and authenticate each other, thus reducing unauthorized access. Understanding MAC addresses helps with troubleshooting network connectivity issues at a lower level.

Bandwidth refers to the transmission capacity of a network, measured in bits per second (bps). It represents the amount of data that can be transmitted over a network connection within a given time. Higher bandwidth means faster data transfer rates. Factors affecting bandwidth include the type of network technology, the physical medium used (e.g., fiber optic cable, copper wire), and network congestion. Understanding bandwidth is crucial for determining network performance and planning for future network growth. Sufficient bandwidth is essential for smooth network operation.

Firewalls are network security systems that monitor and control incoming and outgoing network traffic based on predetermined security rules. They act as a barrier between a trusted internal network and an untrusted external network, such as the internet. Firewalls can be hardware or software-based. They inspect network packets, blocking or allowing traffic based on rules configured by network administrators. Firewalls are essential for protecting networks from unauthorized access, malware, and other threats. Properly configured firewalls are crucial for maintaining network security.

Virtual Private Networks (VPNs) create secure connections over public networks, such as the internet. They encrypt data transmitted between a VPN client and a VPN server, protecting it from eavesdropping and unauthorized access. VPNs are frequently used to protect sensitive data when accessing a network remotely, or to provide secure access to a company's internal network for employees working from home. Understanding VPN technology is critical in today's increasingly distributed and remote work environments. VPNs offer a crucial layer of security in online connectivity.

Network security encompasses the policies and practices adopted to protect a network and its data from unauthorized access, use, disclosure, disruption, modification, or destruction. It incorporates various measures, including firewalls, intrusion detection systems, access control lists, and encryption. A comprehensive network security strategy is crucial for protecting sensitive data and ensuring the confidentiality, integrity, and availability of network resources. Network security is an ongoing process requiring vigilance and adaptation to evolving threats. Investing in robust network security is essential for business operations and data protection.

3.4 Cybersecurity Terms

Understanding cybersecurity requires familiarity with its specialized vocabulary. Terms like "malware," encompassing viruses, worms, trojans, and ransomware, represent a crucial starting point. Malware's goal is to damage, disrupt, or gain unauthorized access to systems, often exploiting vulnerabilities in software or hardware. A vulnerability is a weakness in a system's design, implementation, operation, or internal controls that can be exploited by a threat agent. A threat agent is a malicious actor or event that has the potential to cause harm. Understanding these foundational concepts sets the stage for comprehending more complex cybersecurity challenges.

Next, we encounter the concept of "phishing," a social engineering attack where malicious actors attempt to trick individuals into revealing sensitive information like usernames, passwords, or credit card details. These attacks often masquerade as legitimate communications from trusted sources, utilizing deceptive emails, websites, or text messages. Phishing attacks can lead to significant data breaches and financial losses, making awareness and preventative measures critical. Many organizations implement security awareness training to mitigate this pervasive threat. Effective training emphasizes identifying phishing attempts through analyzing email headers, website URLs, and unusual requests for personal information. This proactive approach is significantly more effective than simply relying on technical security measures.

Network security forms another cornerstone of cybersecurity. Firewalls, acting as barriers between trusted and untrusted networks, filter network traffic based on predefined rules. Intrusion detection systems (IDS) and intrusion prevention systems (IPS) monitor network traffic for malicious activity, alerting administrators to suspicious events or automatically blocking them. These systems represent essential layers of defense against various network-based attacks. In addition to these, Virtual Private Networks (VPNs) encrypt data transmitted over public networks, protecting sensitive information from eavesdropping. Understanding how firewalls, IDS/IPS, and VPNs function is crucial for protecting organizational networks and data. Each of these components requires careful configuration and ongoing monitoring to ensure optimal effectiveness. This includes regular updates to security rules and signatures to keep pace with evolving threat landscapes.

Data security involves techniques and technologies to protect data from unauthorized access, use, disclosure, disruption, modification, or destruction. Encryption, a crucial component of data security, transforms readable data into an unreadable format, ensuring confidentiality. Data loss prevention (DLP) solutions identify and prevent sensitive data from leaving the organization's control. Regular data backups are also crucial for business continuity and disaster recovery, ensuring data can be restored in case of system failure or cyberattack.

These measures, working in tandem, create a robust data security posture, capable of protecting sensitive data from various threats. The importance of data security cannot be overstated, given the rising costs of data breaches and regulatory penalties for non-compliance. Organizations must invest in appropriate security controls, regularly test their effectiveness, and continuously adapt to emerging threats.

Finally, incident response planning is paramount. A comprehensive incident response plan outlines the steps an organization will take to prepare for, identify, analyze, contain, eradicate, recover from, and learn from a security incident. This involves defining roles and responsibilities, establishing communication protocols, and creating a detailed playbook for handling various scenarios. Regular testing and updates of the plan are crucial to ensure its effectiveness in a real-world situation. A well-defined incident response plan minimizes the impact of a security incident, helping organizations to recover quickly and efficiently. This reduces potential financial losses, reputational damage, and operational disruptions. Moreover, a robust incident response capability allows for continuous learning and improvement, strengthening an organization's overall security posture. Remember, prevention is key, but a well-rehearsed response plan mitigates the impact when incidents occur, and this is unavoidable.

3.5 AI Terms

Machine learning, a cornerstone of AI, involves algorithms that allow systems to learn from data without explicit programming. These algorithms identify patterns, make predictions, and improve their performance over time, based solely on the input data they are fed. Think of it as teaching a computer to recognize a cat by showing it thousands of pictures of cats—it learns the features that define a cat, like pointed ears and whiskers, without needing a manual definition of "cat." The success of machine learning hinges on the quality and quantity of the training data; insufficient or biased data leads to inaccurate or discriminatory outcomes. This constant iterative improvement of the machine is a key distinction and one of the most exciting aspects of its potential.

Deep learning, a subset of machine learning, utilizes artificial neural networks with multiple layers (hence "deep") to analyze data. These networks are inspired by the structure and function of the human brain, allowing for the processing of complex, unstructured data such as images, audio, and text. Deep learning has been instrumental in breakthroughs in areas like image recognition, natural language processing, and speech recognition, achieving levels of accuracy previously unimaginable. This complexity, however, comes with increased computational demands and the need for substantial datasets. The deeper the network, the more data it typically needs to be properly trained, and this presents challenges in resources.

Natural language processing (NLP) focuses on enabling computers to understand, interpret, and generate human language. This involves tasks such as machine translation, text summarization, sentiment analysis, and chatbot development. NLP algorithms analyze the grammatical structure, context, and meaning of text to extract valuable information or create meaningful responses. For example, NLP powers virtual assistants like Siri and Alexa, allowing them to respond to voice commands and answer questions. Significant challenges remain in addressing nuances in language, sarcasm, and cultural contexts—a significant amount of contextual information is vital to accurate interpretation.

Computer vision, another critical AI field, allows computers to "see" and interpret images and videos. This involves image recognition, object detection, and image segmentation, enabling applications such as autonomous vehicles, medical image analysis, and facial recognition technology. Advancements in deep learning have dramatically improved the accuracy and speed of computer vision systems, but ethical considerations, particularly concerning privacy and bias in algorithms, are paramount. The development of a robust computer vision program is a significant undertaking, and it is an area of intense research and development.

Neural networks, at the heart of many AI systems, are interconnected nodes (neurons) that process and transmit information. They are inspired by the biological neural networks of the human brain, functioning through weighted connections that modify the signal transmission between nodes, constantly evolving from the input. The strength of these connections is adjusted during training, allowing the network to learn from data and improve its performance over time. Different architectures of neural networks exist, each designed for specific tasks, from simple feedforward networks to complex recurrent and convolutional networks, with each having its own strengths and limitations. Understanding the basic principles of neural networks is essential for grasping the core of many AI algorithms.

Robotics, closely intertwined with AI, involves the design, construction, operation, and application of robots. AI plays a crucial role in enabling robots to perceive their environment, make decisions, and perform complex tasks autonomously. This collaboration produces robots capable of navigating unpredictable environments, interacting with humans, and executing intricate maneuvers. However, integrating AI and robotics presents challenges in areas such as sensor integration, control algorithms, and safety. This combined application is already transforming various industries, and it is likely to be a major disruptive technology for years to come.

Reinforcement learning, a type of machine learning, involves training agents to interact with an environment and learn optimal behaviors through trial and error. Agents receive rewards for desirable actions and penalties for undesirable ones, gradually learning to maximize their cumulative reward. Reinforcement learning is particularly useful in applications requiring decision-making in complex, dynamic environments, such as game playing, robotics, and resource management. However, designing effective reward functions and dealing with exploration-exploitation trade-offs can be challenging. The algorithm needs to carefully balance trying new actions (exploration) to improve itself and using actions that have worked well in the past (exploitation).

Explainable AI (XAI) is a rapidly developing field aiming to make AI decision-making more transparent and understandable. This addresses concerns about the "black box" nature of many AI systems, where it's difficult to understand how they arrive at their conclusions. XAI methods aim to provide insights into the reasoning behind AI predictions, helping to build trust and identify potential biases. This increased transparency can significantly impact the adoption and ethical use of AI in various sectors.

Generative AI uses algorithms to create new content, such as text, images, audio, and video. These systems learn from existing data and generate novel outputs that share characteristics with the training data. Generative AI has applications in art, design, music composition, and content creation. However, concerns around copyright and the potential for misuse remain crucial aspects of the ongoing debate.

4 Medical Terminology

Understanding medical terminology is crucial for effective communication within the healthcare field. Precise language prevents misunderstandings that could have serious consequences for patients. Miscommunication can lead to incorrect diagnoses, inappropriate treatments, and, in the worst cases, even death. Therefore, mastering this specialized vocabulary is not merely beneficial—it's essential for the safety and well-being of those in your care.

The human body is incredibly complex, and describing its intricate structures and functions requires a dedicated vocabulary. Think about the difference between "cardiac arrest" and "respiratory distress"—two terms that, while both serious, point to very different underlying problems. One relates to the heart, the other to the lungs. Accurate terminology instantly clarifies the nature of a medical issue, guiding swift and appropriate responses. Using precise anatomical terms ensures that everyone—doctors, nurses, technicians, and other professionals—is on the same page, facilitating seamless cooperation and effective treatment. For example, knowing the difference between the terms "proximal" and "distal" when referring to limbs is fundamental in documenting injuries or describing surgical procedures.

Many common medical terms are derived from Latin or Greek roots. This is because ancient civilizations made significant advancements in medicine, leaving behind a rich vocabulary that forms the base of modern medical terminology.

Learning the building blocks of this vocabulary—prefixes, suffixes, and root words—can help you decode unfamiliar terms. For instance, "cardio" (heart), "pulmo" (lung), and "hepat" (liver) are common roots, while suffixes like "-itis" (inflammation), "-ectomy" (surgical removal), and "-oma" (tumor) indicate the nature of a medical condition or procedure. Understanding these root components allows you to deduce the meaning of complex terms like "hepatitis" (inflammation of the liver) or "appendectomy" (surgical removal of the appendix), quickly building your medical vocabulary.

Diagnosis is as much about precise language as it is about the physical examination itself. Precise diagnosis terms are essential for accurate record-keeping, facilitating communication among medical professionals and providing clarity for patients. The same condition might be described using different but equally accurate terminology depending on the professional involved. A radiologist will use different terms than a surgeon, yet both will need to accurately relay their observations to ensure consistency in treatment and care. Consider the difference between "benign" and "malignant" tumors, a vital distinction for selecting appropriate treatment strategies and providing accurate prognoses. These terms aren't just labels; they carry profound implications for patient care and future treatment plans.

Medical imaging plays a critical role in modern diagnostics, and understanding the terminology associated with these techniques is vital for interpreting results. Terms like "radiograph," "CT scan," "MRI," and "ultrasound" refer to different imaging modalities, each with its strengths and limitations. Additionally, the terminology involved in describing the findings on those images often employs precise anatomical references and nuanced descriptions of tissue appearances. Learning the language used to describe findings on medical images, such as "lesion," "opacity," or "edema," enhances your ability to fully grasp the diagnostic process and appreciate the role of imaging in forming a clinical picture. This is

key for assisting healthcare professionals in making informed decisions.
Beyond the individual terms themselves, understanding the context and nuances of medical language is also essential. The same word might have different meanings depending on the context or specialty. For example, "syndrome" might refer to a set of symptoms in one case and a specific disease in another. This highlights the importance of considering the broader medical picture, the patient's history, and other accompanying signs and symptoms when interpreting medical records or healthcare communication. Paying attention to these subtleties and learning to interpret medical language within its specific context is critical for competent healthcare assistance. This level of understanding will significantly enhance your ability to decipher and contribute meaningfully to medical discussions and documentation.
Learning medical terminology is an ongoing process. It requires consistent effort and a commitment to expanding your knowledge. It might seem daunting, but by breaking down complex terms into their constituent parts and studying the context in which they are used, you can master this specialized vocabulary. Many resources are available to support this learning—medical dictionaries, online resources, and specialized textbooks can help you build a robust understanding of medical terminology. Remember that the purpose of this learning is to improve the quality of patient care by facilitating clear and precise communication in the medical environment. The commitment to mastering this language is not just about memorization, it's about improving safety, and bettering care.

4.1 Anatomy Terms

Understanding the human body requires a firm grasp of anatomical terminology. Precision is paramount; a single misplaced term can lead to significant misunderstandings, especially in a medical context. Therefore, mastering this vocabulary is crucial for clear communication among healthcare professionals and for accurate patient care. We'll explore essential terms, focusing on clarity and avoiding overly technical jargon.

Let's start with directional terms. Superior refers to a structure above another, while inferior indicates a structure below. Anterior denotes the front, and posterior signifies the back. Medial describes a position closer to the midline of the body, whereas lateral indicates a position further from the midline. Proximal describes a location closer to the point of attachment of a limb, while distal refers to a location further from the point of attachment. Superficial means closer to the surface, and deep signifies further from the surface.

Understanding planes of section is equally vital. The sagittal plane divides the body vertically into right and left halves. A midsagittal plane creates equal right and left halves. The frontal (or coronal) plane divides the body vertically into anterior and posterior portions. Finally, the transverse (or horizontal) plane divides the body horizontally into superior and inferior portions. These planes are fundamental to interpreting anatomical images and understanding spatial relationships within the body.

Now, let's delve into regional terms. The axial region includes the head, neck, and trunk. The appendicular region encompasses the limbs—the upper limbs (arms and hands) and the lower limbs (legs and feet). Within these regions, further subdivisions exist. For example, the abdominal region is often further categorized into quadrants or regions to pinpoint locations more precisely. This level of detail is necessary for effective medical communication.

Body cavities are crucial for organizing internal organs. The dorsal cavity protects the central nervous system, encompassing the cranial cavity (housing the brain) and the vertebral canal (housing the spinal cord). The ventral cavity is larger and subdivided into the thoracic cavity (containing the heart and lungs) and the abdominopelvic cavity (containing the digestive organs, urinary system, and reproductive organs). These cavities provide protection and allow for organ movement.

Moving on to specific anatomical structures, we can examine the skeletal system. Terms like cranium, clavicle, sternum, vertebrae, ribs, humerus, radius, ulna, femur, tibia, fibula, and carpals all refer to specific bones. Understanding their locations and functions is key. The muscular system equally boasts a complex terminology, with terms such as biceps brachii, quadriceps femoris, and deltoids defining specific muscles and their locations.

The circulatory system uses equally precise terminology. Arteries, veins, and capillaries represent the vessels through which blood travels, while terms like aorta, vena cava, and pulmonary arteries designate specific arteries and veins. The heart itself is comprised of four chambers, with specific names for each. Knowing these terms will aid in interpreting information about the heart and blood vessels. Furthermore, a vast vocabulary exists to explain the nervous system, its many nerves, and the brain's intricate regions.

Finally, let's consider the digestive system. The esophagus, stomach, small intestine, large intestine, liver, gallbladder, and pancreas, are all integral organs. The respiratory system terms such as the trachea, bronchi, and lungs are essential for understanding breathing mechanisms. Similarly, the urinary system uses specific terms for the kidneys, ureters, bladder, and urethra. Each of these systems requires a unique vocabulary to accurately depict its functions and components. The reproductive system holds yet another significant layer of specialized terminology.

Mastering these terms is essential for anyone working within the medical field or studying anatomy. Precise language prevents miscommunication and ensures everyone is on the same page, leading to safer and more efficient care. Further research and continued use of these terms will solidify your understanding and improve your ability to communicate effectively within the medical and scientific communities. The more familiar you become with these anatomical terms, the greater your comprehension of the human body's intricate structure and functions. Remember to consult anatomical atlases and medical dictionaries for additional information and visual aids. Consistent review and application are vital to retain this specialized vocabulary.

4.2 Diagnosis Terms

Medical diagnoses rely heavily on precise terminology. A seemingly minor difference in wording can drastically alter the understanding of a patient's condition, influencing treatment decisions and overall prognosis. Accuracy is paramount, not just for doctors communicating amongst themselves but also for effectively conveying information to patients and their families. Clear and unambiguous language ensures everyone is on the same page, reducing misunderstandings and fostering trust. This shared understanding minimizes the risk of errors and promotes better patient outcomes, regardless of the specialty or specific ailment. Effective communication, driven by precision in language, forms the cornerstone of successful medical care. Failing to use appropriate terms can lead to misdiagnosis and ineffective treatment.

Consider, for instance, the difference between "congestive heart failure" and "cardiomyopathy." While both relate to heart function, the underlying causes and treatment strategies differ significantly. One describes a condition where the heart struggles to pump blood effectively, while the other encompasses a range of diseases affecting the heart muscle. Accurate diagnosis hinges on this level of detail. Similarly, differentiating between "bronchitis" and "pneumonia" requires careful assessment of symptoms and diagnostic tests. The former involves inflammation of the bronchi, whereas the latter is a lung infection, demanding specific interventions to ensure effective recovery. Mislabeling can delay the appropriate treatment or lead to ineffective interventions, endangering the patient's wellbeing.

The language surrounding diagnoses frequently employs abbreviations and acronyms, adding another layer of complexity. Understanding these shortcuts is crucial for medical professionals. For example, "CHF" commonly signifies congestive heart failure, while "CAD" represents coronary artery disease. Familiarity with such abbreviations speeds up communication and allows for efficient record-keeping. However, using such abbreviations without sufficient context can lead to mistakes. Therefore, clinicians must always ensure clarity and avoid ambiguous interpretations when conveying critical information regarding a patient's condition. Accurate documentation is vital, particularly in the medical field, to prevent misunderstandings.

Medical imaging plays a significant role in diagnosis. Terms describing the different imaging modalities, like "MRI," "CT scan," "X-ray," and "ultrasound," are common in daily medical practice. Each technique offers unique insights into the body's internal structures, and understanding their respective capabilities is essential for interpreting results effectively. Furthermore, the descriptions of findings within those images often require specific vocabulary. Terms like "lesion," "mass," "nodule," and "cyst," all require a careful approach in their interpretation. Context within the image, the patient's history and other findings from the physical examination must be considered. A clear and complete understanding of the terms used will lead to better decision-making and patient care.

Differential diagnosis, a process of identifying a specific disease from a list of possibilities, involves careful consideration of symptoms, medical history and test results. This process demands a nuanced vocabulary enabling healthcare providers to articulate reasons for favoring one diagnosis over another. The accurate use of specific terms in the documentation of such processes prevents future misinterpretations or disputes related to the diagnosis, treatment, or overall management of the case. Accurate and precise language prevents misunderstandings and facilitates collaborative efforts among medical professionals, ultimately benefiting the patient. The use of detailed and descriptive language regarding the decision-making process aids in preventing medical errors and malpractice claims.

5 Legal Terminology

Understanding legal terminology is crucial, not just for lawyers, but for anyone navigating the complexities of our justice system. A seemingly simple phrase can hold immense weight, dictating outcomes and shaping legal battles. Mastering this language isn't just about memorizing words; it's about grasping the underlying principles and nuances that define each term's meaning.

Consider, for instance, the term "contract." It seems straightforward enough, but the legal definition is significantly more nuanced. It requires an offer, acceptance, and consideration—an exchange of something of value. Breach of contract, a common legal claim, hinges on proving a failure to fulfill these obligations. The specific implications differ based on the contract type; whether it's written or oral, express or implied, influences the interpretation and potential legal repercussions considerably. Understanding these nuances is key to crafting sound contracts and avoiding disputes.

Criminal law presents a completely different landscape of terminology. You'll encounter terms like "actus reus," the guilty act, and "mens rea," the guilty mind. These elements are fundamental to establishing criminal liability. Proving both is necessary for a successful prosecution, and the lack of either can lead to acquittal. Beyond these core concepts, you'll encounter specific offenses, each with its own detailed requirements and potential penalties. Terms like manslaughter and murder, seemingly similar, possess vast legal differences rooted in intent, circumstances, and premeditation. Understanding these distinctions is critical for anyone involved in criminal proceedings.

Civil law, in contrast to criminal law, focuses on disputes between individuals or entities. Here, terms like "negligence," "tort," and "damages" become paramount. Negligence involves a failure to exercise reasonable care, leading to harm. A tort is a civil wrong, encompassing actions like negligence, defamation, and trespass. Damages are the monetary compensation awarded to the injured party. The specifics of these terms are far-reaching, influencing everything from evidence gathering to the determination of liability. Understanding the nuances of civil law empowers you to protect your rights and interests in non-criminal disputes.

Beyond the core elements of these three main branches, legal terminology extends to procedural aspects. Terms like "plea bargain," "discovery," and "voir dire" are essential for understanding how cases progress through the legal system. A plea bargain involves negotiations between the prosecution and defense, often leading to a reduced sentence in exchange for a guilty plea. Discovery is the process of gathering evidence before trial, crucial for both sides to build their cases effectively. Voir dire is the jury selection process, where potential jurors are questioned to ensure impartiality. Each process relies heavily on detailed understanding and mastery of relevant legal terminology.

In the sphere of international law, the complexities multiply exponentially. Concepts like "jurisdiction," "extraterritoriality," and "treaty" demand even deeper comprehension of the legal frameworks involved. Navigating international legal matters requires a sophisticated understanding of varying legal systems and an ability to translate legal concepts across borders. The very structure of legal argumentation, built upon defined terms and precedents, necessitates precise knowledge. Misunderstanding a single term can have significant, and potentially devastating, consequences.

Finally, remember that legal terminology is constantly evolving. New laws, precedents, and societal shifts influence its meaning and application. Staying informed and updated is essential for anyone seeking to navigate this complex, ever-changing field. Continuous learning and engagement with legal resources are indispensable for maintaining a grasp of the dynamic nature of legal language. Therefore, building a solid foundation in legal terminology is a continuing process, requiring constant updating and refinement, to ensure continued competency. Its mastery is not a destination, but a journey of ongoing learning and adaptation.

5.1 Contract Law

A contract, fundamentally, is a legally binding agreement between two or more parties. This agreement creates mutual obligations, enforceable by law. Understanding the core components—offer, acceptance, and consideration—is paramount. An offer is a clear and definite proposal, indicating a willingness to enter into a contract. Acceptance signifies unequivocal agreement to the terms of the offer, forming the basis of the contract. Consideration refers to something of value exchanged between parties; it's the mutual benefit each party receives. Without consideration, the agreement lacks legal force.

The intricacies of contract law extend far beyond these basic elements. For instance, capacity refers to the legal ability of parties to enter into a binding agreement. Minors, those deemed mentally incapacitated, or intoxicated individuals often lack the necessary capacity. Similarly, duress or undue influence, situations where one party forces or manipulates another, can render a contract voidable. These elements highlight the necessity of ensuring all parties involved are fully aware and voluntarily consenting.

Furthermore, the principle of privity of contract dictates that only parties to the agreement can sue or be sued on it. Third parties typically lack standing to enforce the contract, unless specifically provided for within its terms. This crucial aspect significantly limits who can engage in legal action concerning the contract's fulfillment or breach. It emphasizes the importance of carefully considering all involved parties and their rights from the outset.

Mistakes, misrepresentations, and illegality further complicate the contractual landscape. A mutual mistake, where both parties are mistaken about a fundamental aspect of the agreement, can invalidate the contract. Misrepresentation, a false statement of fact that induces the other party to enter the contract, can lead to legal repercussions, potentially rendering the contract voidable. Contracts involving illegal activities are, naturally, unenforceable. Thorough due diligence and clear communication are therefore crucial in avoiding these pitfalls.

Terms within contracts are classified as either express or implied. Express terms are explicitly stated, either orally or in writing. Implied terms are not explicitly stated but are inferred by law or custom. Understanding this distinction is critical for interpreting contractual obligations. The interpretation of contractual terms often involves analyzing the language used, the parties' intentions, and relevant case law. Ambiguity can lead to disputes, underscoring the importance of precise and unambiguous language.

Breach of contract occurs when one party fails to perform their contractual obligations. Remedies for breach vary and depend on the specifics of the situation. These can range from monetary damages (compensatory, punitive, or nominal) to specific performance, where a court orders the breaching party to fulfill their obligations. Injunctive relief, prohibiting a party from performing a certain action, may also be an available remedy. The selection of an appropriate remedy hinges on the severity of the breach and the desired outcome.

Finally, understanding the termination of contracts is equally vital. Contracts can be terminated by performance, where both parties fulfill their obligations;

agreement, where the parties mutually consent to terminate; frustration, where an unforeseen event renders performance impossible; or breach, as already discussed. The method of termination significantly affects the parties' rights and obligations. Careful consideration of potential termination scenarios should be included in the contract's formation. Therefore, consulting legal counsel during the creation and interpretation of contracts is strongly advised to ensure clarity, compliance, and protection of rights for all parties involved.

5.2 Criminal Law

Criminal law revolves around acts deemed harmful to society, punishable by the state. Understanding its terminology is crucial for legal professionals, students, and anyone interacting with the justice system. The precise use of words can mean the difference between conviction and acquittal, freedom and incarceration.

Precise definitions matter significantly. Consider "theft," seemingly straightforward. However, the nuances—larceny, embezzlement, robbery, burglary—each carry distinct legal meanings and potential penalties. The slightest mischaracterization could alter the course of a case. This necessitates careful study and understanding of legal definitions, often found in statutes and case law. The subtle differences between these words are often the source of complex arguments in courts, demonstrating why precise language is absolutely fundamental.

The concept of mens rea, or "guilty mind," illustrates the importance of mental state in criminal liability. Unlike strict liability offenses, most crimes require proof of intent, recklessness, or negligence, specific elements that must be proven beyond a reasonable doubt. The absence of mens rea, demonstrated through effective defense strategies, could lead to an acquittal. This highlights the intricate interplay between factual evidence and legal definitions within the criminal justice system.

The term actus reus, meaning "guilty act," completes the foundational elements of most crimes. It demands a demonstration of a voluntary physical act, which when combined with the necessary mens rea, constitutes a criminal offense. However, the definition of "voluntary" can be contentious, with complexities arising from issues like duress, automatism, or intoxication. Again, careful interpretation is required. The prosecution must therefore provide evidence and demonstrate that the act was voluntary and intentional, connecting the act directly to the accused's guilt.

Furthermore, understanding specific offenses is vital. Murder, manslaughter, assault, battery—these words represent distinct levels of culpability, requiring different evidentiary standards and carrying vastly different sentences. A single act could fall under multiple legal categories, depending on the circumstances, highlighting the complexity inherent in criminal law. The application of criminal law frequently involves the careful consideration of the context of the crime, taking into account factors such as provocation, self-defense, and the circumstances surrounding the event.

Beyond the core offenses, numerous procedural terms define the legal process. Arrest, indictment, arraignment, plea bargaining, trial, sentencing—each carries weight and impact on the accused's rights and the outcome of the case. Knowing the meaning of these words is critical in navigating the complex pathways of the criminal justice system, regardless of your direct involvement. Even the smallest aspect of the legal process requires a complete understanding of its terminology.

Finally, understanding the terminology surrounding defenses is critical. Self-defense, insanity, duress, and alibi each provide a legal basis to negate criminal liability, demanding precise application. Successfully claiming a defense often depends on a thorough understanding of the legal standards, requiring a deeper analysis of applicable statutes and precedent. Failure to understand this terminology can easily result in unsuccessful court arguments and unfavorable outcomes. Therefore, the study of the specific terminology of these defenses requires a comprehensive understanding of the legal system.

Therefore, mastering criminal law terminology goes beyond simple memorization. It involves understanding the nuances of definitions, the interplay between elements of offenses, and the procedural intricacies of the system. This level of comprehension is critical for clear communication, effective legal representation, and overall justice. The precision inherent in criminal law terminology underscores its importance in safeguarding both individual rights and the interests of society. The entire judicial system operates on a foundation of defined terminology, and those who seek to understand the legal processes and outcome must be proficient in its use. The mastery of this vocabulary is paramount for accuracy and fairness.

5.3 Civil Law

Civil law encompasses a broad range of legal disputes between individuals, organizations, or both. Unlike criminal law, which focuses on offenses against the state, civil law addresses private wrongs and seeks to provide remedies for harm suffered. These remedies often involve monetary compensation or injunctive relief, designed to restore the injured party to their pre-harm position as much as possible. The complexities inherent in civil litigation necessitate a deep understanding of specific terminology.

Negligence, a cornerstone concept in civil law, involves a failure to exercise the care that a reasonably prudent person would exercise in similar circumstances. Establishing negligence requires proving four elements: duty of care, breach of duty, causation, and damages. Duty of care means a legal obligation to avoid causing harm to others. A breach occurs when the defendant fails to meet this obligation. Causation links the defendant's actions directly to the plaintiff's injuries, and damages represent the actual harm suffered by the plaintiff, whether physical, emotional, or financial. The specifics of proving each element can be incredibly nuanced, often relying on expert witness testimony and intricate analysis of evidence.

Contracts form another significant area within civil law. A legally binding contract requires offer, acceptance, consideration, and an intention to create legal relations. An offer is a clear expression of willingness to enter into an agreement, while acceptance signifies unqualified agreement to the terms of the offer. Consideration represents the value exchanged by each party – it can be money, goods, services, or a promise to do or refrain from doing something. The intention to create legal relations demonstrates that both parties intended their agreement to be legally enforceable. Failure to meet these conditions can result in a contract being deemed void or unenforceable, leaving one or both parties without legal recourse. Disputes regarding contract interpretation, breach of contract, or the validity of a contract itself are frequently litigated in civil courts.

Torts, civil wrongs that result in harm to another, constitute a broad category encompassing a variety of actions. These include negligence, discussed above, but also intentional torts such as assault, battery, defamation, and trespass. Intentional torts require proof of intentional conduct causing harm, whereas negligence requires proof of carelessness. Defamation, for instance, involves the publication of false statements that harm the reputation of another. To succeed in a defamation claim, the plaintiff must show that the statement was false, published to a third party, and caused them actual damage. The complexities of proving these elements, particularly the existence of malice, often lead to protracted and intricate litigation.

Property law, an essential facet of civil law, addresses the rights and obligations associated with ownership and possession of real and personal property. Real property encompasses land and anything permanently attached to it, while personal property includes movable possessions. Property disputes can arise from issues such as ownership, boundaries, easements, and leases. Establishing clear title to property, determining the extent of ownership rights, and enforcing property agreements often requires specialized legal knowledge and expertise, particularly in matters of land registration and conveyancing. The complexities of property law extend to various forms of ownership, including joint tenancy, tenancy in common, and trusts, each carrying its own set of rights and responsibilities.

Family law, although often treated separately, falls under the broader umbrella of civil law. It governs matters related to marriage, divorce, child custody, and support. Dissolving a marriage involves intricate legal procedures, often involving the division of marital assets, child support arrangements, and the determination of custody and visitation rights. These proceedings demand a nuanced understanding of legal precedents, case law, and statutory frameworks which vary widely based on jurisdiction. Determining child custody, in particular, requires careful consideration of the best interests of the child, a complex determination that relies on a wide range of factors.

Understanding the key terminology within these specific areas of civil law is paramount. Each concept, from negligence to contract formation, possesses its own precise legal meaning and carries specific implications for legal outcomes. Mastering this precise language is crucial not only for legal professionals but also for anyone involved in civil disputes, enabling clearer communication and a more informed approach to navigating complex legal situations. The subtleties within each area require thorough research and comprehension, highlighting the importance of accurate and precise terminology.

6 Business Terminology

Understanding financial statements is crucial for any business professional. Terms like assets, representing what a company owns, and liabilities, signifying what it owes, are foundational. Equally important are equity, representing the owners' stake, and the interconnectedness of these elements within the fundamental accounting equation: Assets = Liabilities + Equity. Proficiency in analyzing these statements—the balance sheet, income statement, and cash flow statement—allows for informed decision-making regarding investment, budgeting, and financial health assessment. Mastering these concepts unlocks a deeper comprehension of a company's financial standing, enabling effective strategic planning.

Moving beyond the balance sheet, a grasp of key financial ratios is essential. These ratios offer a comparative analysis of a company's performance over time or against industry benchmarks. For example, liquidity ratios, such as the current ratio, assess a company's ability to meet its short-term obligations. Solvency ratios, like the debt-to-equity ratio, evaluate its long-term financial stability and risk profile. Understanding and calculating these ratios provides crucial insight into a company's financial strength and its capacity to withstand economic fluctuations. Effective analysis of these ratios underpins informed investment decisions and ensures proactive risk management.

The world of marketing is equally rich in specialized vocabulary. Market segmentation, the process of dividing a broad consumer market into sub-groups, allows for targeted marketing strategies. Understanding consumer behavior, including motivations and purchasing patterns, is fundamental to effective market research. Branding, encompassing the creation of a unique identity and perception of a product or company, is pivotal to market success. A well-defined brand strengthens market position, fosters customer loyalty, and differentiates from competitors. Therefore, understanding these concepts is critical for establishing a strong presence and attracting the intended customer base.

Effective marketing relies heavily on various promotional strategies. Advertising uses paid media to reach a target audience. Public relations, or PR, focuses on building positive relationships with various stakeholders, aiming for favorable media coverage and enhancing reputation. Digital marketing, a vital component in today's landscape, uses online channels for promotion, including search engine optimization (SEO), social media marketing, and content marketing. Understanding and integrating these different promotional strategies is crucial for achieving optimal marketing outcomes and aligning communication with the overall marketing goals.

Another crucial aspect of business terminology is understanding different business structures. A sole proprietorship, owned and run by one person, has simple setup but unlimited liability. A partnership, involving two or more individuals, shares profits and losses but also carries joint and several liability. Limited liability companies (LLCs) offer the advantage of limited liability, separating personal assets from business debts. Corporations, characterized by complex legal structures, have limited liability for shareholders but face more stringent regulatory requirements. Choosing the right structure impacts taxes, liability, and administrative burdens, significantly affecting the operational dynamics and future development of the business.

Finally, consider the terms related to business transactions. Sales revenue represents the income generated from sales activities. Cost of goods sold (COGS) accounts for the direct costs associated with producing goods sold. Gross profit, calculated as sales revenue minus COGS, reveals the profitability of goods or services before deducting other expenses. Net profit, the bottom line after considering all expenses, reflects the overall profitability of the business. Understanding these core terms facilitates accurate financial reporting, aids in cost management and pricing strategies, and is indispensable for making informed decisions about the overall financial trajectory of a business. Fluency in these terms is, therefore, fundamental to business success.

6.1 Finance Terms

Understanding financial statements requires familiarity with specific terminology. A balance sheet, for instance, presents a snapshot of a company's assets, liabilities, and equity at a specific point in time. Assets are what a company owns, such as cash, accounts receivable (money owed to the company), and property, plant, and equipment (PP&E). Liabilities represent what a company owes to others, including accounts payable (money owed to suppliers), loans, and other debts. Equity is the residual interest in the assets of an entity after deducting its liabilities; it essentially represents the owners' stake in the company. Analyzing these components provides insights into a company's financial health and stability.

The income statement, conversely, shows a company's financial performance over a period of time, typically a quarter or a year. It details revenues (money earned from sales), cost of goods sold (COGS), gross profit (revenue minus COGS), operating expenses (costs associated with running the business), and net income (profit after all expenses are deducted). Understanding the relationship between revenues, expenses, and profit is crucial for assessing a company's profitability and growth potential. Key ratios derived from these statements, such as gross profit margin and net profit margin, offer valuable comparative metrics for assessing performance relative to industry benchmarks or prior periods. Understanding these ratios helps to analyze the profitability of a company over time.

Cash flow statements are just as important, providing details on the movement of cash both into and out of a company during a specific period. They are broken down into operating activities (cash flows from the main business operations), investing activities (cash flows related to investments in assets), and financing activities (cash flows related to borrowing, debt repayment, and equity financing). Analyzing the cash flow statement allows you to understand the sources and uses of a company's cash, offering a different perspective than the income statement's accrual accounting approach. A healthy cash flow is essential for a company's short-term survival and long-term growth. Ignoring this aspect can lead to erroneous assessments of the company's true financial strength.

Beyond the basic financial statements, several key financial ratios provide further insight into a company's performance. Liquidity ratios measure a company's ability to meet its short-term obligations; examples include the current ratio (current assets divided by current liabilities) and the quick ratio (a more stringent measure excluding inventory). Solvency ratios assess a company's ability to meet its long-term debt obligations, including the debt-to-equity ratio (total debt divided by total equity) and the times interest earned ratio (earnings before interest and taxes divided by interest expense). Profitability ratios, such as return on assets (ROA) and return on equity (ROE), indicate how effectively a company is using its assets and equity to generate profit. Understanding and calculating these financial ratios is critical for a complete financial assessment.

Investing in securities necessitates understanding various terms. For example, stocks represent ownership in a company, providing shareholders with voting rights and the potential for dividends. Bonds are debt instruments issued by corporations or governments, promising to repay the principal amount plus interest at a specified future date. Mutual funds pool money from multiple investors to invest in a diversified portfolio of securities, offering a convenient way to access a broader market. Exchange-traded funds (ETFs) are similar to mutual funds but trade on stock exchanges like individual stocks, offering greater flexibility and transparency. Each investment vehicle carries unique risks and rewards, requiring careful consideration of personal risk tolerance and financial goals before making an investment decision. Understanding these instruments and the terminology associated with them is paramount for success in the financial markets.

6.2 Marketing Terms

Market research is the systematic gathering, recording, and analysis of data about consumers, competitors, and the market itself. It's the bedrock of any successful marketing campaign, providing insights into consumer preferences, buying habits, and market trends. Understanding your target audience is crucial, and research helps you define their demographics, psychographics, and buying behaviors—information vital for crafting effective marketing messages. Without this foundational understanding, your marketing efforts will likely be ineffective, wasting resources and failing to connect with your potential customer base. This process involves various methods such as surveys, focus groups, and data analysis, each offering unique advantages and helping paint a clearer picture of your market. Careful planning and interpretation are crucial for effective market research and achieving marketing objectives.
Branding is the process of creating a unique identity for your product or service that differentiates it from competitors. It encompasses the name, logo, visual elements, messaging, and overall experience associated with your offering. Think of a strong brand as a promise to your customers – a promise of quality,

value, and consistency. A well-defined brand builds trust, loyalty, and recognition, making it easier to attract and retain customers. To build a powerful brand, you must articulate your brand's unique selling proposition (USP) and consistently communicate that proposition across all touch points. It's not just about creating a logo; it's about building a comprehensive brand strategy that guides all aspects of your marketing and business operations. Neglecting brand building can make it difficult to compete, even with superior products.

Advertising encompasses the paid promotion of products or services through various media channels. Unlike public relations, which relies on earned media, advertising involves direct payment for exposure. Effective advertising requires careful planning and execution, considering the target audience, messaging, and media selection. Common advertising channels include television, radio, print, online advertising, and social media marketing. The key is to craft compelling messages that resonate with your target audience and encourage them to take action—whether it's making a purchase, visiting a website, or learning more about your product. The success of advertising is often measured by key performance indicators (KPIs) such as click-through rates, conversion rates, and return on investment (ROI). A well-executed advertising campaign enhances brand awareness, drives sales, and generates leads.

A marketing strategy is a comprehensive plan that outlines your marketing objectives, target audience, and the tactics you'll use to achieve your goals. It's not merely a collection of individual marketing activities but a cohesive framework guiding all your marketing efforts. A robust marketing strategy starts with clearly defined goals – what do you want to achieve with your marketing? Increased brand awareness? Higher sales? Lead generation? Once you have defined your objectives, you can develop a plan to achieve them, incorporating tactics such as content marketing, search engine optimization (SEO), social media marketing, email marketing, and paid advertising. Regular monitoring and analysis of your marketing efforts are crucial to identify what's working and make adjustments along the way. Without a clear strategy, your marketing efforts are likely to be fragmented and ineffective.

Customer relationship management (CRM) involves managing interactions with current and potential customers. It's about building and maintaining strong relationships that foster loyalty and advocacy. Modern CRM systems often utilize software to track customer interactions, manage sales leads, and personalize marketing communications. The focus is on understanding customer needs and preferences, providing exceptional customer service, and using data to drive better business decisions. CRM strategies contribute to increased customer retention, improved customer satisfaction, and ultimately, higher profitability. Understanding your customer lifecycle is a key component of effective CRM, enabling you to tailor your communications and offerings at every stage of the customer journey. Ignoring customer relationship management can lead to a high customer churn rate and a failure to build brand loyalty.

Marketing automation involves the use of software tools to automate repetitive marketing tasks, improving efficiency and consistency. This can include email marketing campaigns, social media scheduling, lead nurturing, and customer segmentation. The goal is to streamline marketing processes, allowing marketers to focus on higher-level strategic initiatives. Effective automation strategies personalize marketing communications, improving the customer experience and driving better results. Implementing marketing automation requires careful planning and selecting appropriate tools that integrate seamlessly with your existing systems. Marketing automation tools free up valuable time for more creative and analytical marketing work. Failing to automate repetitive tasks can lead to burnout and reduced efficiency, hindering the company's ability to meet marketing goals.

7 Appendix: Resources

This appendix offers a curated selection of resources to further your exploration of terminology across diverse fields. It's designed to supplement the information within this book and provide you with avenues for deeper learning and specialized knowledge acquisition. The resources included represent a starting point, a springboard for your own dedicated research. Remember that the ongoing evolution of language means new terms and resources will continuously emerge, so staying updated through active engagement is vital. Finding the right resource often means understanding your own learning style and needs.

Begin with online dictionaries. Many specialize in specific terminologies, catering to professionals and students alike. For example, specialized medical dictionaries provide detailed definitions and etymologies for medical terms, enabling a more thorough understanding of their origins and precise meanings, far beyond what a general dictionary might offer. Similarly, legal databases often incorporate detailed case law summaries along with precise definitions, helping to disambiguate complex legal concepts. In the technological realm, online documentation for various software and hardware platforms contains vast terminology glossaries.

Beyond online dictionaries, several excellent books delve into the intricacies of specific terminologies. For legal professionals, treatises on contract law or criminal procedure offer meticulously defined terms within their respective contexts.

Similarly, in the scientific realm, textbooks on physics, chemistry, or biology not only explain core concepts but also meticulously define the related terminology. Seek out publications written by reputable academics or professional organizations for the highest standards of accuracy. Remember to cross-reference information found in multiple sources to ensure clarity and understanding. Don't hesitate to explore books dedicated to the history of terminology within specific fields. This approach can enhance your comprehension of current terminology by providing context for its evolution and its nuances.

Don't overlook the power of specialized glossaries and acronym lists. Many professional organizations and companies publish these resources, often available on their websites or as downloadable PDF documents. These glossaries provide quick access to definitions and explanations, invaluable for deciphering complex jargon. They can act as invaluable checkpoints, reinforcing comprehension and resolving ambiguity when encountering unfamiliar acronyms. Remember to always verify the authority and reliability of any source. Explore reputable professional organizations and academic institutions, always verifying the credibility of the source.

In your pursuit of knowledge, think critically and engage actively. Don't just passively absorb information, instead actively question and cross-reference definitions. Compare and contrast terminology from different sources to identify nuances and potential areas of ambiguity. Constructing your own personal glossary as you learn is a valuable strategy that will reinforce understanding and assist in future research. Moreover, consider seeking out professional associations or societies related to your area of interest, as these organizations are excellent resources for the most up-to-date terminology in those specific fields and sometimes even offer specialized training or educational materials.

Developing a strong grasp of terminology is a continuous process of learning and refinement. It requires consistent effort, a critical eye, and a commitment to engaging with diverse resources. This appendix, then, isn't just a list; it's a pathway toward a deeper understanding of the world's intricate languages, a journey of intellectual growth, and a key to unlocking deeper comprehension within any chosen discipline. Remember that the most effective use of these resources involves careful consideration of the context in which terminology is used. By combining these resources with consistent learning, you will cultivate a superior understanding of terminology in any domain. By consistently applying critical thought, you will hone your skills and become more adept at navigating the complex landscape of specialized terminology. The pursuit of terminological expertise is a valuable endeavor. Therefore, approach it with determination and the commitment to lifelong learning.

7.1 Further Reading

This chapter isn't about finding more terms, but about where to find them when you need them. Think of it as your roadmap to deeper understanding, a guide to continuing your journey into the world of precise language. You've learned the basics; now it's time to explore the deeper waters.

A good starting point would be specialized dictionaries. While general dictionaries provide a foundation, subject-specific dictionaries are invaluable for detailed definitions and nuanced meanings. Look for dictionaries focused on your area of interest; for example, if you're studying international law, a legal dictionary dedicated to international law is far superior to a general legal dictionary. These resources often offer detailed explanations, historical context, and examples within the relevant field. These specialized resources are often found in academic libraries or online through subscription databases.

Consider exploring professional journals and publications. These resources aren't just repositories of research, but also excellent sources for learning the current terminology. By reading articles and studies within a specific field, you'll encounter terms used in context, strengthening your comprehension and understanding of subtle distinctions between seemingly similar words. The writing style in these journals will hone your ability to decipher complex vocabulary and grasp intricate concepts. Keep an eye out for introductory articles and reviews – these often provide a helpful overview of important terminology for newcomers.

Don't underestimate the power of textbooks. Well-written textbooks, especially those used in university courses, are meticulously crafted to explain concepts clearly and consistently. These books not only provide definitions, but also place those definitions within a broader conceptual framework. Textbook language is often carefully chosen to avoid ambiguity and ensure a high degree of clarity. Pay attention to the glossaries included at the end of most textbooks; these are invaluable for quick reference and reinforcement.

Another valuable resource, often overlooked, is online encyclopedias. While the quality can vary widely, reputable online encyclopedias (like those published by university presses or established academic institutions) often offer concise yet informative explanations of key terminology. These resources provide a broad overview that can then be supplemented by more focused research.

However, be critical. Always cross-reference information from multiple sources. Different authorities may have slightly different definitions or preferences for terminology. Comparing multiple perspectives ensures a well-rounded understanding, safeguarding you against relying on potentially biased or incomplete information. This process of comparing and contrasting information enhances your understanding beyond simply memorizing definitions.

Finally, remember that terminology evolves. New terms emerge, and existing terms develop new connotations. Staying abreast of changes within your field is essential to maintaining precise and up-to-date understanding. Regularly consulting updated dictionaries and journals is crucial; it's a continuous process of learning and refinement. The ongoing nature of terminology is a reminder of the dynamic and ever-changing nature of knowledge and professional language. Your commitment to continuous learning will allow you to master and advance in your chosen field. The pursuit of precision is a lifelong journey, one enriched by constant exploration and diligent research. Embrace this ongoing process, and the power of precise language will be your constant companion. Always seek further resources for the best information.

7.2 Online Dictionaries

Online dictionaries offer a powerful tool for navigating the intricacies of specialized terminology. They're not simply repositories of definitions; they're dynamic resources that can significantly enhance comprehension and precision in professional and scholarly work. Accessing these dictionaries efficiently, however, requires understanding their strengths and limitations.

Many online dictionaries cater to specific fields, providing far more nuanced definitions than general-purpose resources. Consider, for instance, a legal professional needing to understand the exact meaning of "consideration" in contract law. A general dictionary might offer a broad, everyday interpretation, while a dedicated legal dictionary would delve into the precise legal ramifications and precedents surrounding that term. This level of specificity is crucial for accurate legal work. Similarly, a medical professional would benefit from a medical dictionary which provides in-depth explanations of anatomical terms or diagnostic procedures, far beyond what a layman's dictionary could provide.

Finding the right dictionary for your needs involves careful consideration of the field and level of detail required. Begin by identifying the subject area; a search for "online medical dictionary" or "legal terminology dictionary" yields specialized results, offering definitions that resonate with those working within those specific fields. Review multiple resources; different dictionaries might emphasize particular aspects or adopt different terminological conventions. Comparing their entries can give a fuller understanding of the complexities inherent in specialized vocabularies.

Beyond simple definitions, many online dictionaries provide additional valuable information. Some dictionaries incorporate etymologies—tracing the historical evolution of words— which can illuminate the nuances of a term and its connotations. Others link terms to related concepts, creating a rich network of interconnected meanings. Such cross-referencing can significantly aid in developing a deep understanding of a subject area. Moreover, some dictionaries include illustrative examples, demonstrating the appropriate usage of terms in context, which is far more instructive than simply memorizing definitions.

However, relying solely on online dictionaries can present challenges. Not all online dictionaries maintain equal standards of accuracy or scholarly rigor. Some may reflect colloquial usages rather than established professional definitions. Furthermore, the rapidly evolving nature of terminology, especially in technology and science, can mean that definitions in some dictionaries become outdated quickly. Therefore, it's critical to consult dictionaries that have reputable editorial oversight and frequent updates. Looking for dictionaries associated with established academic institutions or professional organizations can give you a higher degree of confidence in their authority.

Supplementing online dictionaries with other resources is always recommended. Standard textbooks and scholarly articles often provide context and deeper explanations of key terms. Consulting multiple sources helps to form a more comprehensive understanding. Actively engaging with the terminology through writing, discussions, and application in your own work allows for a true understanding of its intricacies and practical implications.

Ultimately, the effective use of online dictionaries is about strategic resourcefulness. It's about recognizing their limitations as well as their power. By understanding the specific strengths of different online dictionaries, employing appropriate search strategies, and always critically evaluating the information presented, you can leverage these resources to enhance your understanding and communication within any specialized field of study or practice. The meticulous and informed use of these tools contributes significantly to clarity of expression and precision of thought in your work, minimizing ambiguity and maximizing impact. Remember, precision in language is paramount in any professional field.

Remember also that the constantly evolving nature of language necessitates continuous engagement with these resources. Regularly consulting relevant online dictionaries allows you to stay abreast of any changes or updates to terminology and definitions, ensuring accuracy and relevance in your work. This proactive approach is fundamental for any professional seeking to maintain a high standard of expertise.

Finally, cultivate the habit of actively questioning terminology. Don't simply accept definitions passively. Consider the context in which a term is used, its relationship to other terms, and its implications within the broader field of study. By engaging in this critical evaluation process, you move beyond rote memorization and develop a deep, insightful understanding of specialized terminology, transforming passive knowledge into active expertise. This practice transforms dictionary consultation from a simple lookup to an ongoing process of learning and refinement.

7.3 Glossary of Acronyms

This chapter provides a glossary of acronyms frequently encountered across various fields covered in this book. Understanding these abbreviations is crucial for efficient navigation of specialized literature and effective communication within professional contexts. We've organized the entries alphabetically for easy reference.

A.I.: Artificial Intelligence. This refers to the simulation of human intelligence processes by machines, particularly computer systems. These processes include learning (acquisition of information and rules for using the information), reasoning (using rules to reach approximate or definite conclusions), and self-correction. A.I. applications range from simple expert systems to complex deep learning algorithms, constantly evolving with advancements in computing power and data availability. The field is broad and interdisciplinary, drawing from computer science, mathematics, linguistics, psychology, and neuroscience. Furthermore, ethical considerations surrounding A.I. development and deployment are increasingly prominent.

CAD: Computer-Aided Design. This is the use of computer systems to assist in the creation, modification, analysis, or optimization of a design. CAD software provides tools for creating 2D and 3D models, simulating physical properties, and generating manufacturing instructions. It's widely used in various industries, from architecture and engineering to automotive design and manufacturing. The precision and efficiency of CAD significantly impact design processes, allowing for intricate details and rapid prototyping. Moreover, CAD facilitates collaboration among design teams and improves the accuracy of final products.

CEO: Chief Executive Officer. This is the highest-ranking executive in a company, responsible for overall management and strategic direction. The CEO typically reports to the board of directors and is accountable for the company's performance and profitability. The role involves leadership, decision-making, strategic planning, and overseeing all aspects of the organization's operations. Furthermore, CEOs often represent the company publicly and interact with stakeholders, including investors, customers, and employees. The responsibilities and scope of a CEO can vary considerably depending on the size and structure of the organization.

DBMS: Database Management System. This refers to software used to create, maintain, and access databases. A DBMS provides tools for data definition, data manipulation, and data security. It facilitates efficient storage, retrieval, and manipulation of large volumes of structured data. Popular DBMS examples include MySQL, PostgreSQL, Oracle Database, and Microsoft SQL Server. The choice of DBMS depends on factors such as the size and type of data, scalability requirements, and budget. Effective database management is crucial for organizations that rely on data for decision-making and operational efficiency.
DNA: Deoxyribonucleic Acid. This is a molecule that carries the genetic instructions for all living organisms. It's a double helix structure composed of nucleotides, each containing a deoxyribose sugar, a phosphate group, and a nitrogenous base (adenine, guanine, cytosine, or thymine). The sequence of these bases determines the genetic code, which dictates the synthesis of proteins and regulates cellular functions. The study of DNA has revolutionized biology and medicine, leading to advancements in genetic engineering, disease diagnosis, and personalized medicine. Furthermore, DNA analysis is used in forensic science and evolutionary biology.
GDP: Gross Domestic Product. This is the total monetary or market value of all the finished goods and services produced within a country's borders in a specific time period. GDP is a key indicator of a country's economic performance and is used to measure economic growth. Calculating GDP typically involves summing the value added at each stage of production. Different methods exist for calculating GDP, including the expenditure approach, the income approach, and the production approach. Understanding GDP trends is vital for policymakers and investors to gauge economic health and make informed decisions. Fluctuations in GDP can signal periods of economic expansion or contraction.
IoT: Internet of Things. This refers to the network of physical objects—"things"—embedded with sensors, software, and other technologies for

the purpose of connecting and exchanging data with other devices and systems over the internet. IoT devices can range from wearable fitness trackers to industrial sensors and smart home appliances. The interconnectedness of IoT devices facilitates data collection, automation, and remote monitoring. Security concerns and data privacy are significant challenges in the rapidly growing IoT landscape. The potential applications of IoT are vast and span various industries, impacting how we live, work, and interact with our environment.

NMR: Nuclear Magnetic Resonance. This is a physical phenomenon involving the absorption and re-emission of electromagnetic radiation by atomic nuclei in a magnetic field. It's used extensively in various scientific fields, particularly in chemistry and medicine. In chemistry, NMR spectroscopy provides information about the structure and dynamics of molecules. In medicine, magnetic resonance imaging (MRI) uses NMR principles to create detailed images of the human body's internal structures. The applications of NMR extend to materials science, environmental science, and food science, making it a versatile analytical technique. Advancements in NMR technology continue to improve its sensitivity and resolution.

RAM: Random Access Memory. This is a type of computer memory that can be accessed randomly; that is, any byte of memory can be accessed directly without accessing other memory locations. RAM is volatile memory, meaning that its contents are lost when the power is turned off. It's used to store data and instructions that are currently being used by the computer's processor. The amount of RAM in a computer significantly impacts its performance; more RAM allows for faster processing of information. Different types of RAM exist, each with varying speeds and capacities. Selecting appropriate RAM is crucial for optimizing computer performance.

ROI: Return on Investment. This is a financial metric that measures the profitability of an investment relative to its cost. It's calculated by dividing the net profit from an investment by the cost of the investment. ROI is expressed as a percentage and is used to evaluate the efficiency and effectiveness of different investment options. High ROI indicates a successful investment, while low or negative ROI suggests that the investment has not generated sufficient returns. Companies and investors use ROI to assess the financial viability of various projects and make informed decisions about resource allocation. It's an essential tool for evaluating the financial performance of investments.

www.ingramcontent.com/pod-product-compliance
Lightning Source LLC
LaVergne TN
LVHW022025200926
848481LV00007B/2544